WIRING REPAIRS

Many household electrical projects are simple and can be done safely as long as you follow some basic rules. Replacing a switch, an outlet, or a light fixture are all straightforward repairs that can improve the function or appearance of your home. There are situations, however, when you should seek the help of a licensed electrician or electrical contractor. Safety, of course, is paramount. If you aren't sure that the power is off to a circuit; if you don't understand what you see in an electrical box or don't know what to do next; or if the wiring is old (cloth and rubber insulated, for instance), nongrounded, or aluminum, call in a pro.

The cable used in most houses since the early 1970s is nonmetallic sheathed cable with a grounding wire (commonly known as NM or Romex®, which is one brand). That's the focus of this booklet. The other types of wiring—knob-and-tube, armored cable (BX), or wire in conduit—can be complex to work on and will not be covered here.

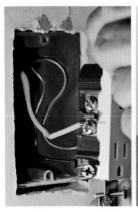 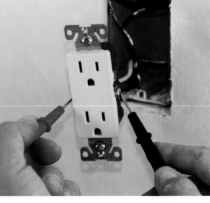 

EASY REPAIR. Replacing a receptacle is a simple task, as long as you follow some basic safety rules (here, testing for voltage).

SAFETY FIRST

For any home wiring project, you must shut off the power to the circuit you're going to work on. Make no mistake, common household voltage (120 volts) can seriously injure or kill. Shut off power to the circuit at the main panel or the subpanel, and verify by testing that it's off *every time* before you handle wiring.

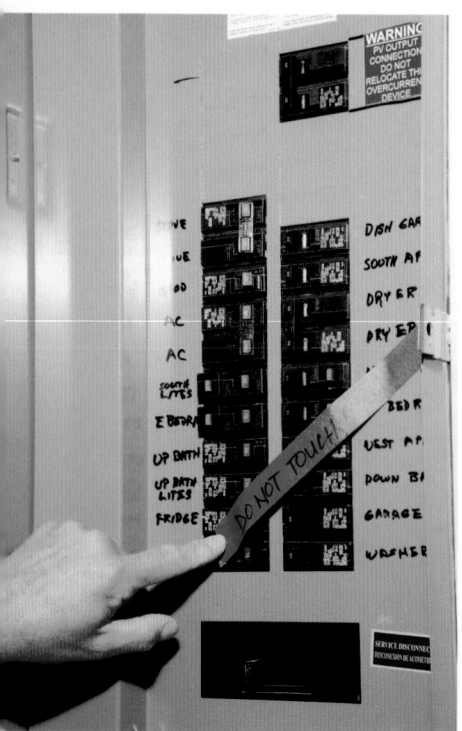

QUICK TIP It's important to understand the circuit you're working on; a fixture could be fed by a timer switch or a photocell, and if power is not cut at the breaker, wires you're working on could become energized, shocking or electrocuting you. If in doubt, cut power to the whole house, and even then, test before you work.

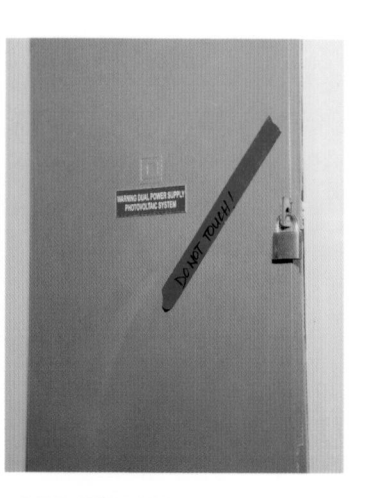

LEAVE A CLEAR MESSAGE. When you shut off the breaker feeding the circuit you're working on, use tape to "tag out" the panel so that no one turns the power back on while you're working.

BEST APPROACH—LOCK IT! Use a padlock to "lock out" the panel for full safety (and be sure to hold onto the key yourself).

TESTING FOR VOLTAGE

There are two ways to test for voltage after shutting off the power: using a noncontact voltage tester (a "volt-tick") or a voltage-continuity tester. Each tester has strengths and limitations, but I recommend using both for better safety as there are certain situations where one tester or the other will give false-negative results. Using both testers only takes a minute and could save your life.

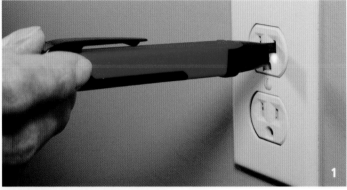

QUICK TIP Understand and follow the manufacturer's instructions, and always check your tester on a live circuit before and after using it. Most testers are powered by batteries—and batteries do go dead.

1. CHECK AT THE FACE. Insert the tip of the noncontact voltage tester (top) into each slot. It will light or beep if voltage is present. Do a second check with the voltage-continuity tester (left) to verify the results before opening up the box.

2. CHECK IN THE BOX. There may be a second circuit passing through the box, so touch the tester to as many wires as possible.

3. CHECK THE TERMINAL SCREWS. If there is a broken part inside the receptacle, this test could save your life.

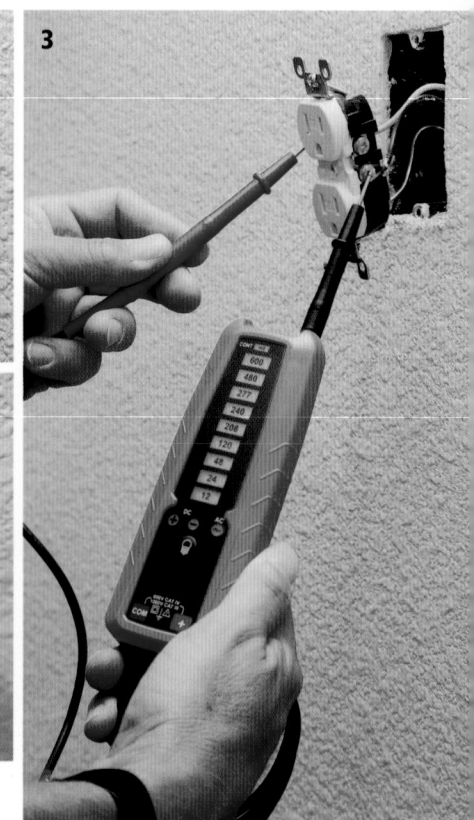

4. CHECKING FOR PROPER WIRING. A plug-in polarity tester can tell you if the wires are connected to the right terminals. The three lights come on in various combinations to indicate correct or various incorrect wiring.

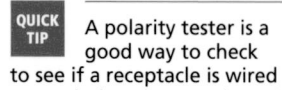

> **QUICK TIP** A polarity tester is a good way to check to see if a receptacle is wired correctly, but do not rely on it to tell you if power is off.

WHAT YOU'LL NEED

You won't need a lot of expensive tools to do simple electrical work, but there are a few special electrician's tools that you need to do quality work. These tools include the two testers shown on p. 3, side-cut or lineman's pliers (used to twist wires to make connections), and wire strippers to remove the plastic insulation from wires without damaging them. A crimper properly crimps connecting sleeves of soft copper for ground-wire connections. Some lineman's pliers also have crimpers built in.

These specialty tools, and typical homeowner tools like screwdrivers, a utility knife, and a hammer, will serve for electrical work.

CUT AND TWIST WITH LINEMAN'S PLIERS. Make sure to get a pair that fits your hand.

PREP WIRE WITH WIRE STRIPPERS. This took can be used to remove wire insulation and also to cut up to five sizes of bolts.

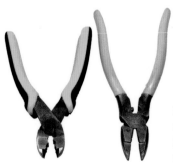

MAKE CONNECTIONS WITH CRIMPERS. Cut and crimp with the pair on the left, or use the crimp pocket on the lineman's pliers on the right.

BASIC SKILLS

There are a number of basic skills you'll need to perform reliable and durable work.

STRIPPING WIRE

Nonmetallic cable has a plastic sheath or jacket enclosing insulated wires and a bare ground wire. The insulation must be removed from the wires to make connections. Use a good-quality wire stripper for that.

QUICK TIP Make sure to use the right hole. If it's too small, you damage the wire; too large, and it'll be hard to pull off the slug of insulation.

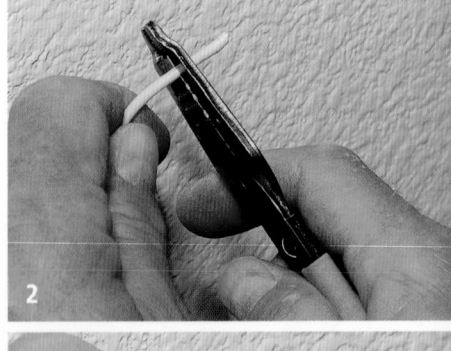

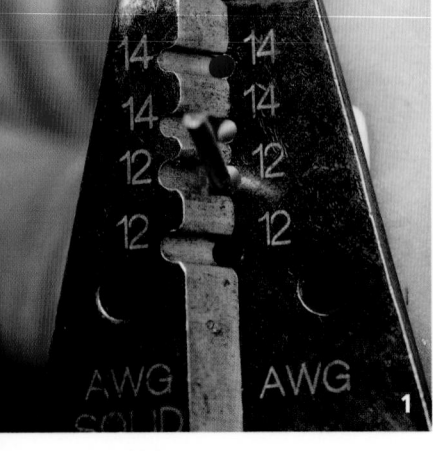

1. CHECK WIRE SIZE. Check the gauge (diameter) of the wire with the stripper hole. The wire shown here is 14 gauge and is clearly too small for the 12-gauge hole.

2. STRIP THE WIRE. When you've found the right gauge hole, squeeze the stripper closed, release just a bit, and use one of the stripper jaws to slide the insulation slug off of the wire end.

MAKING A LOOP

To attach the wire to a switch or receptacle, you need to loop the wire end so it can attach under the screw terminals. To loop the wire, use the looping hole in the stripper or a pair of long-nose pliers.

1. CATCH THE END. Insert the stripped wire end into the small hole on the face of the stripper.

2. TWIST AND LOOP. Flip your wrist 180 degrees and bend a loop in the wire.

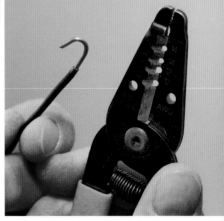

QUICK TIP The wire in cables comes in different gauges or sizes. Most home wiring is either 14 gauge or 12 gauge; 14 gauge is smaller in diameter than 12 gauge and can carry less current than 12 gauge. In some circuits, like a bathroom receptacle or kitchen counter receptacles, electrical code requirements mandate the heavier 12-gauge wire.

STANLEY

CONNECTING THE WIRE TO THE TERMINAL

Place the wire loop over the terminal screw so that tightening the screw will close the loop. Tighten the screw, making sure that the wire loop is oriented in a clockwise direction so that the screw will close the loop as the screw is tightened and is under the screw for one-half to two-thirds of the screw.

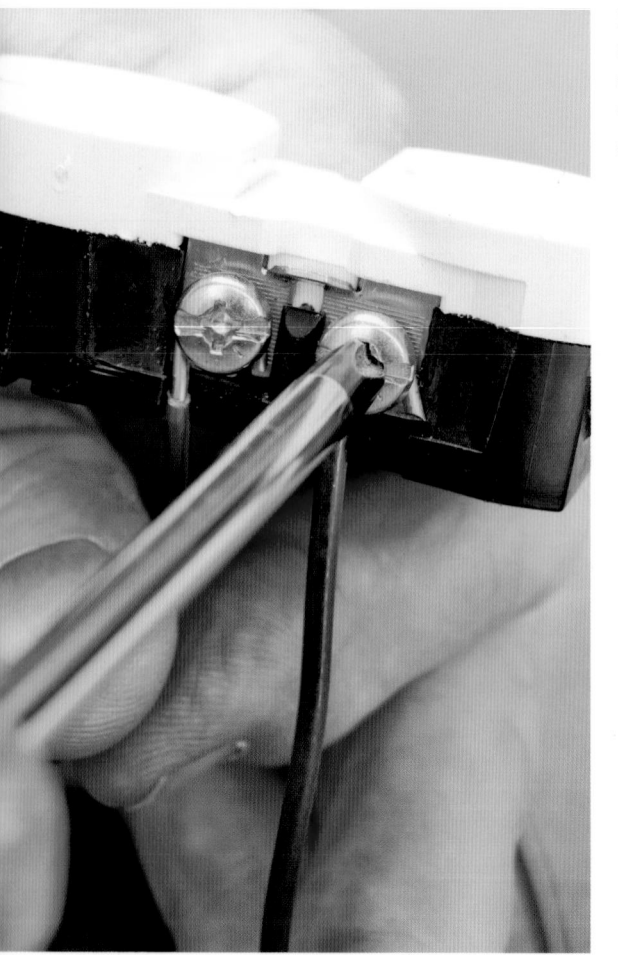

MAKE THE CONNECTION. Place the wire loop so that the screw closes the loop.

QUICK TIP Make sure there's no wire insulation caught under the screw head as this could prevent the screw from making good contact with the wire.

MAKING A SPLICE

At many receptacles, at every switch, and at light fixtures, segments of wire need to be connected to each other. These wire-to-wire connections are called splices. To make a splice, strip wires about ¾ in., line them up, twist together, and attach a connector.

1. STARTING A SPLICE. Line up the three stripped wires and grab with lineman's pliers.

2. TWIST TOGETHER CLOCK-WISE. Make a solid connection by twisting enough to put two twists in the insulated part of the wires.

3. CUT THE ENDS EVEN. Use lineman's pliers to trim the end of the twisted wire; a long wire could pierce the tip of the connector.

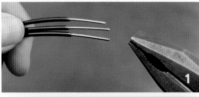

4. SPIN ON A CONNECTOR. Turn the connector clockwise over the stripped wires.

5. COVER IT ALL. The connector skirt must completely cover the stripped parts of the wires.

REPLACING A RECEPTACLE OUTLET

Receptacles and switches are the electrical devices that get the most use in the house, and they do wear out. You'll know a receptacle needs replacing if it's cracked, or if the plug isn't held tightly in the receptacle. Over time, cheap receptacles that are well used simply lose their grip. Other reasons to replace an outlet might be a desire to change the color or style, or to install a ground-fault circuit interrupter (GFCI) receptacle ("shock stopper") or a tamper-resistant ("childproof") receptacle for safety.

QUICK TIP Cheap standard receptacles (and most switches) have "push-in" terminals, where a spring inside grabs the wire. These can get weak and fail sooner or later. Avoid using this method of connecting a wire to a receptacle.

RECEPTACLES DON'T LAST FOREVER. A cracked outlet should be replaced as soon as possible before it fails and causes a shock or a fire.

CHOICES ABOUND. Left to right below, a standard receptacle; a heavier-duty, specification-grade receptacle; and a GFCI receptacle. The upgrade to specification grade is well worth it. GFCIs are required by Code in bathrooms, kitchens, garages, outside, and other locations. All styles come in different colors.

QUICK TIP In almost every instance in a home, it's OK to use a 15-amp-rated receptacle or GCFI. A special 20-amp-rated receptacle is needed only when an appliance or tool that uses more than 15 amps (and so equipped with a special plug) is going to be plugged in.

REMOVING THE OLD RECEPTACLE

Replacing a receptacle is a straightforward repair, and safe—as long as you first cut power at the breaker. Test for voltage to verify that the power is off, using both volt-tick and voltage-continuity testers, and then disconnect the receptacle. Label the wires. Figure out the wiring configuration (see the drawings on pp. 12–13).

TURN OFF THE POWER AT THE BREAKER. Don't trust the labels; verify by testing that you've got the right breaker.

QUICK TIP There may be a second circuit, live, in any box, so always test for voltage even if the power to the receptacle has been shut off.

Receptacle Wiring Configurations

END OF RUN

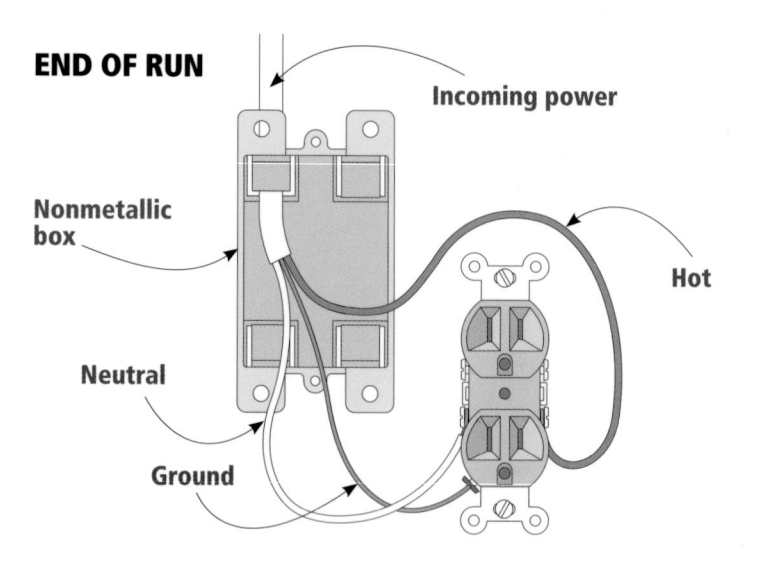

Incoming power

Nonmetallic box

Hot

Neutral

Ground

MIDDLE OF RUN

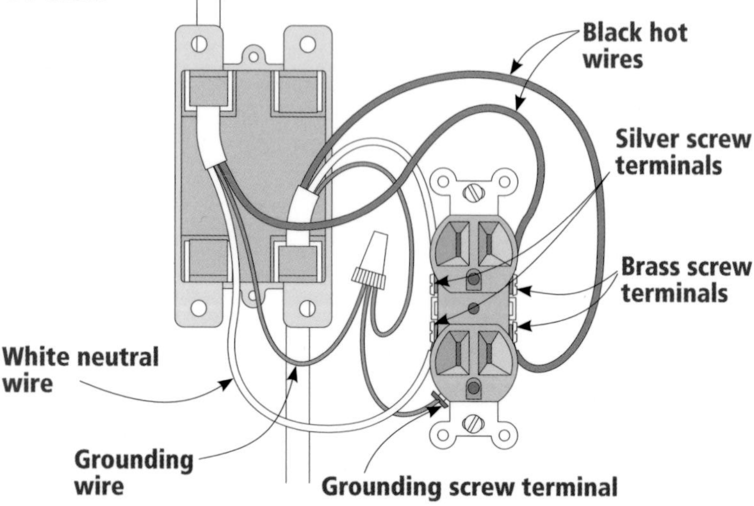

Black hot wires

Silver screw terminals

Brass screw terminals

White neutral wire

Grounding wire

Grounding screw terminal

SPLIT OUTLET

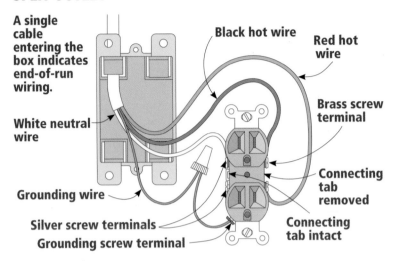

A single cable entering the box indicates end-of-run wiring.

Black hot wire

Red hot wire

White neutral wire

Brass screw terminal

Connecting tab removed

Grounding wire

Connecting tab intact

Silver screw terminals

Grounding screw terminal

Check for Damaged Wire Insulation

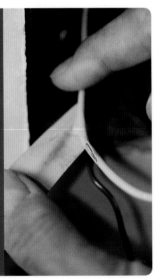

Sometimes when a razor knife is used to remove the cable sheath or jacket, the wire insulation is cut. The bare wire is unsafe and must be fixed before the new receptacle is installed. Wrap two layers of tape on the diagonal over damaged areas, overlapping half the width of the tape. Cut the tape to length rather than tearing it, and you'll get a longer-lasting repair.

WRAP AND COVER. If the insulation on a wire is nicked or torn, wrap it with high-quality electrical tape. Use tape that's the color of the wire insulation.

"Push-in" Backwire Terminals

■ If you're replacing a receptacle with "push-in" terminals, you have two options for disconnecting the wires. If the wires are long enough, just cut them off flush with the back of the device. If they're short, use a very small screwdriver to release them.

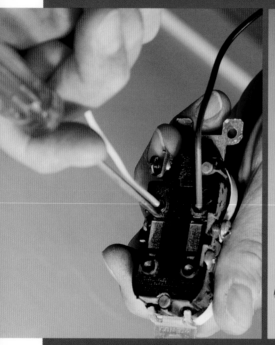

PUSH-IN TERMINALS. The wire is held by a small spring, which may loosen over time.

PUSH AND RELEASE. Insert a thin, flat-blade screwdriver in the hole next to the wire. At the same time, pull on the wire to release it.

INSTALLING THE NEW RECEPTACLE

As explained in "Basic Skills," strip about ¾ in. of insulation from each wire, form a loop, place it under the terminal screw, and tighten the screw. If there are two black or two white wires, you can place one on each screw. By convention, black wires are the "hot" or supply wires and go to the brass-colored screws; the white wires are the "neutral" or return wires and go to the silver-colored screws. The green-insulated or bare copper wire(s) are the equipment grounding wire(s), which are key to safety for tools and appliances with three-prong plugs. The ground wire connects to the grounding screw, which is typically green. If there's more than one ground wire, you must splice them together and include a short jumper wire (called a "pigtail") to connect to the single terminal of a switch or receptacle.

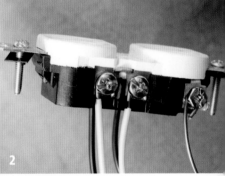

2. CONNECTING TWO CABLES. With two cables in a box (for a middle-of-run circuit), use the device terminals to make the connections. One cable brings power into the box and is connected to the receptacle; the other cable, also connected to the receptacle and so to the first cable, carries power to the next device in the circuit. ▷

1. LOOP AND TIGHTEN. Orient the loop on the terminal screw so that tightening the screw clockwise closes the loop.

3. CONNECTING THREE CABLES. In a receptacle box with three cables, you splice the three black wires together with a pigtail, which connects to the receptacle terminal. Do the same with the white wires and twist on the connectors, which need to be good and tight.

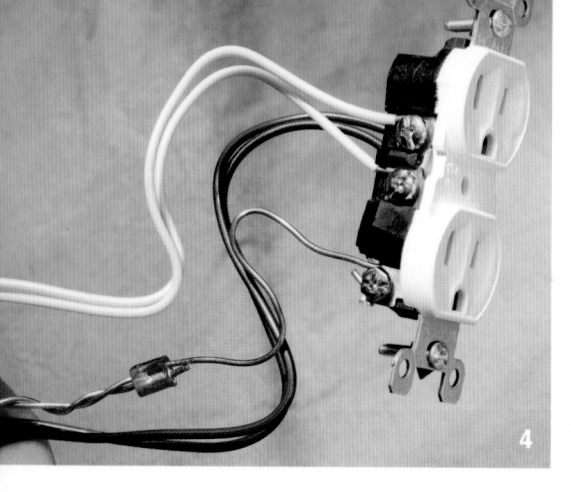

QUICK TIP Make sure the bare ground wire(s) are tucked to the side and bottom of the box, away from the outlet terminal screws.

4. TUCK IN THE WIRES. Fold the wires back into the box, making a Z so that the wires accordion back in neatly.

5. SCREW THE RECEPTACLE INTO THE BOX. Run one screw in half way, then the other, and then finish off both screws.

6. CHECK ALIGNMENT. Make sure the receptacle is vertical, either by eye or with a level.

QUICK TIP When everything is done, closed up, and powered up, use a plug-in tester (see p. 4) to check polarity. If there's a problem, cut power and check your work. If you can't identify the problem, call an electrician.

7. INSTALL THE COVER PLATE. Set the screws so the slots are vertical. It just looks better that way.

GFCI Receptacles

■ Ground-fault circuit interrupter receptacles (GFCIs or GFIs) shut off (interrupt) power if there's even a minute amount of power leaking from the circuit through some unintended path. This will protect you from a bad shock or electrocution if, because of a broken or frayed wire, the power is leaking through you.

GFCI RECEPTACLES (or GFCI circuit breakers) are required on circuits that serve receptacles in the kitchen, bathroom, garage, outside, and other locations. These locations present the greatest risk of someone using electrical items when they're in contact with water or the ground. These GFCI outlets (and better-quality regular outlets) have clamp-type terminals to make connections. These are easy to wire and make very good connections. You strip the insulation off the wire, insert it into the hole, and tighten a screw to clamp the wire tight.

WIRING A GFCI RECEPTACLE. GFCIs have two sets of terminals on the back: One is marked "LINE" (for incoming wires) and the other is marked "LOAD." (for outgoing wires that need GFCI protection, like those going to other kitchen counter outlets). The stripped neutral of the cable providing power fits into the hole next to the silver-colored screw. When the silver screw is tightened, a clamp inside securely grips the wire. The arrangement is the same for the black wire.

REPLACING A SWITCH

There are a number of reasons you might need to replace a switch. The switch might be broken, the light may not come on reliably because the switch is failing or has a loose connection, or you might simply want to change the switch's color or style. As with receptacles, there are a number of different switch options available.

STANDARD TOGGLE. This simple, single-pole switch (below) is the most commonly installed. It is identifiable by its three screw terminals: two brass-colored screws and a green grounding screw.

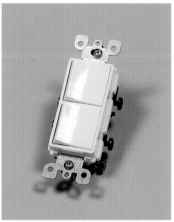

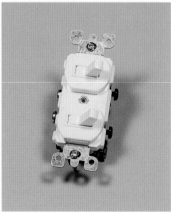

OTHER OPTIONS. A stack switch (right photos) fits two switches into the space of one. The more modern and sleek decorator-style switch is at top; the standard toggle, at bottom.

CONTROL A LIGHT FROM MORE THAN ONE LOCATION. Two three-way switches allow switching from two locations (the toggle style is shown above, but the switches are also available in decorator style).

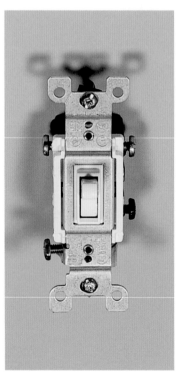

AUTOMATIC SWITCHES SAVE ENERGY. A timer (countdown) switch (below left) is used for the bath vent fan, while a vacancy sensor (bottom left) shuts off the power when no motion is detected in a room.

HOW TO REPLACE A SWITCH

To replace a switch, first cut the power at the breaker and verify that the power is off by observing the light go off when the breaker is shut off. Take off the cover plate, check with a volt-tick to be sure that there is no voltage present, remove the switch mounting screws, and pull the switch out. Test again to be sure the power is off at the switch terminals and in all wires in the box. Sometimes there are two different circuits in a box.

1. CHECK FOR VOLTAGE. Before removing the switch, insert the tip of the tester at the sides of the switch; this checks for a loose wire that may still be connected to power. After pulling the switch out, check for voltage again.

Wiring a Single-Pole Switch

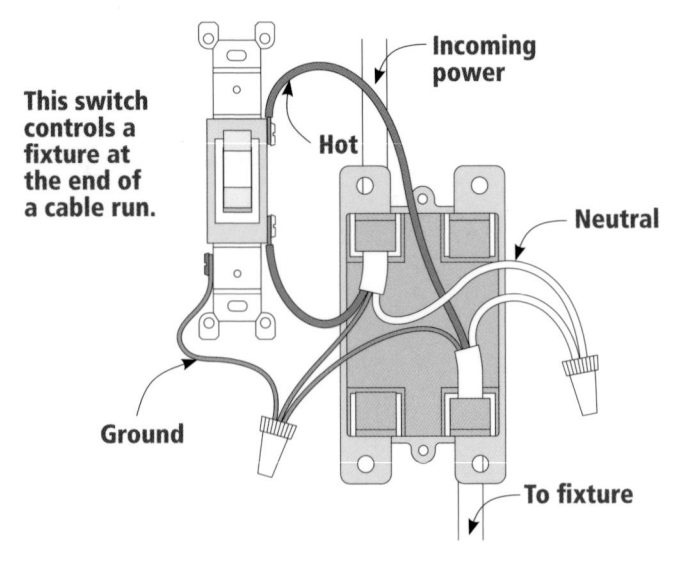

This switch controls a fixture at the end of a cable run.

Incoming power

Hot

Neutral

Ground

To fixture

Look at the switch you are replacing. If it has two brass screws and a green screw, it's a simple switch, also called a single-pole, single-throw switch (the drawing above shows the wiring configuration for a single-pole switch). If it has one black and two brass screws and a ground, it's a three-way switch (a single-pole, double-throw switch). If it has four brass screws and a green screw, it's a four-way switch. For a three-way switch, mark the wire attached to the black screw, using a bit of tape so that you can be sure to connect this wire to the black screw on the new switch. Because of the complexity of four-way switches, I suggest you call an electrician to replace a four-way switch.

Remove the wires from the existing switch one at a time, and connect them to the appropriate terminal of the new switch. The two black wires go to the brass-colored terminal screws, and the bare copper (or green in-sulated) ground wire goes to the green ground screw on the switch. Fold the wires into the box as described previously, screw the switch to the box, and install the cover plate. Power up the circuit and test.

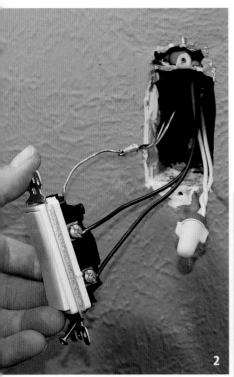

QUICK TIP Simple switches have a mark to indicate which end of the switch is the top, or "up," so that when the handle is down, the switch is off. Make sure to install the switch in the correct orientation. Three-way and four-way switches don't have a top, so they can be installed in any orientation.

2. A SIMPLE SWITCH WITH CONNECTIONS MADE. The two black wires go to the brass terminals, and the bare copper wire is connected to the ground screw.

QUICK TIP If the wires from the old switch are too short to install the new switch (or outlet), add a length of wire. Use a twist-on wire connector to make a splice.

3. THREE-WAY SWITCH WIRED IN. As marked before from the original switch, one wire (either power in, or the wire to the light fixture) goes to the black terminal, two wires go to the brass screws, and the bare copper wire (not visible here) goes to the ground screw.

REPLACING A LIGHT FIXTURE

Basic options for light fixtures include ceiling fixtures and wall fixtures (or sconces). A fixture may have a screw base for a standard incandescent light-bulb socket or a pin base for a compact fluorescent bulb.

To replace a fixture, shut off the power and verify by testing that it's off. Remove the old fixture. Take the shade off, then take the bulbs out. There are usually two screws that hold the fixture base onto a mounting bracket. There will be three wires from the fixture to the building wiring in the ceiling (or wall): a black, a white, and a bare ground, each spliced to the small fixture wires of the corresponding color.

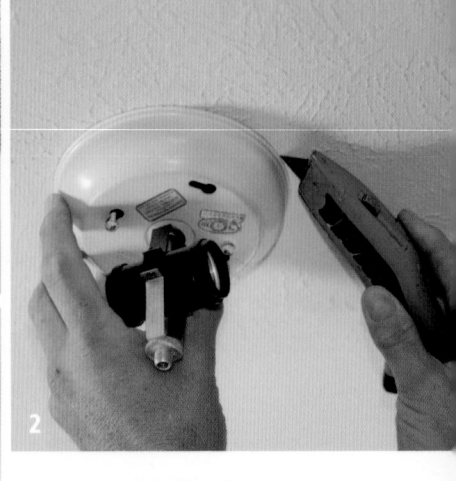

1. REMOVE THE SHADE. For some fixtures, the glass shade rotates to release; for others, there's a decorative nut on the bottom of the shade.

2. BREAK THE PAINT. If the fixture base is stuck to the ceiling with paint, use a utility knife to score the paint to prevent paint chipping around the fixture.

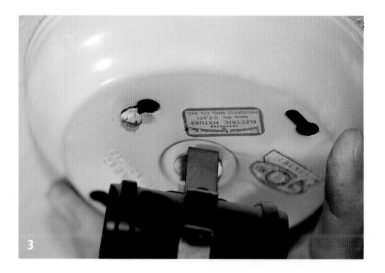

3. LOOSEN THE MOUNTING SCREWS.
Some fixtures have keyhole-shaped holes for mounting. Loosen the screws slightly and rotate the fixture to dismount it.

Test for voltage on each wire. Cut the fixture wires on the fixture side of the splice, then check the wires in the electrical box for voltage. Remove the old mounting plate.

QUICK TIP When removing a fixture, often the screws can be loosened and the fixture rotated a bit so that the screw heads will clear wider openings in the mounting holes. This is much easier than completely removing the two screws.

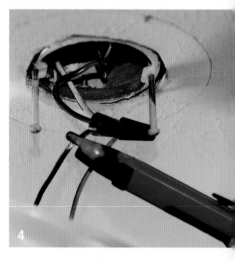

4. CHECK EACH WIRE FOR VOLTAGE.
Use a volt-tick to verify that the circuit is off and safe to work on. Here, the glowing red light indicates that the circuit is still on.

Wiring a Light Fixture

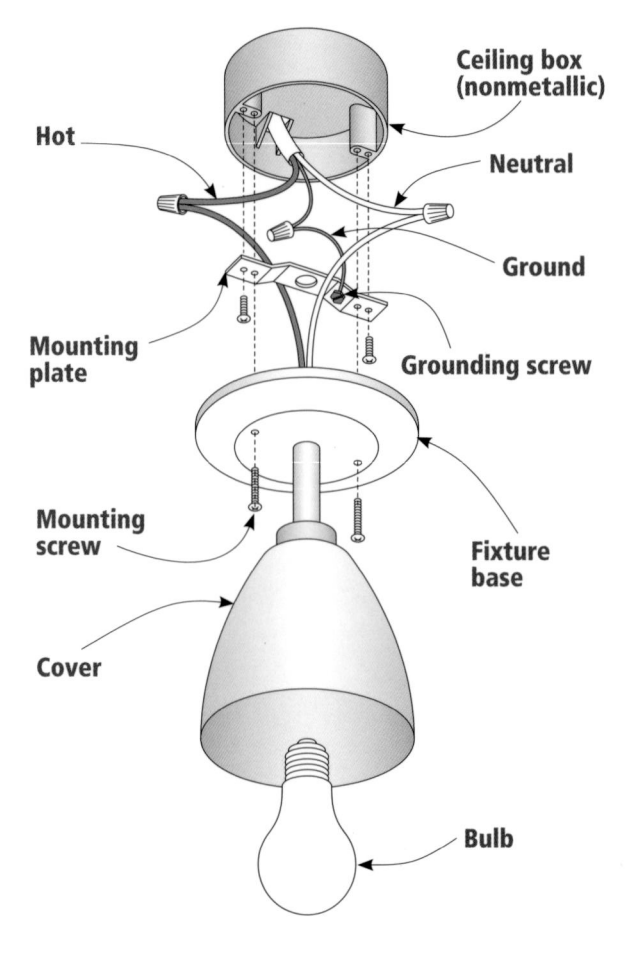

Hot

Ceiling box (nonmetallic)

Neutral

Ground

Mounting plate

Grounding screw

Mounting screw

Fixture base

Cover

Bulb

INSTALL THE NEW MOUNTING BRACKET. If there are two screws used to mount the fixture base to the mounting plate, screw them into the mounting plate before you attach the fixture base to the ceiling box (if there are keyhole mounting slots).

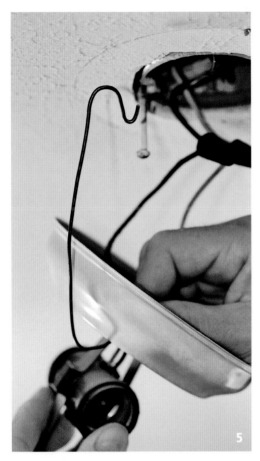

5. HOOK AND HOLD.

Don't let a fixture hang on any of the electrical wiring, as this could damage the wiring. Instead, use a hook made of heavy coat-hanger wire to hold the fixture while you make the splices (see p. 28); attach the top end of the hook to the mounting bracket.

> **QUICK TIP** If you want to replace a light fixture with a ceiling fan, make sure the electrical box is strong enough to handle the weight and the motion of the fan. This usually means replacing the existing ceiling box with one that is "ceiling-fan rated."

Tuck the splices and wires back into the electrical box, and finish by installing the fixture base, then the bulbs, then the shade. Power up the circuit and test. If the lights don't go on, turn off the power at the switch, and remove and replace the bulb(s). Sometimes new sockets are stiff and the bulb doesn't seat all the way. If that doesn't do it, replace the bulbs; sometimes there's a bad bulb or two in with the fixture. If that doesn't do it, cut power to the circuit, and check the splices. That'll usually get things working.

Splicing Small Wires to Bigger Wires

■ The trick to connecting a small, stranded fixture wire to a larger, solid building wire is to twist the stranded wire around the solid wire and leave a bit of the stranded wire past the solid wire. This way, the wire connector will grab the solid wire first. After you've twisted on the connector, give the stranded wire a gentle tug to make sure it's been grabbed by the connector.

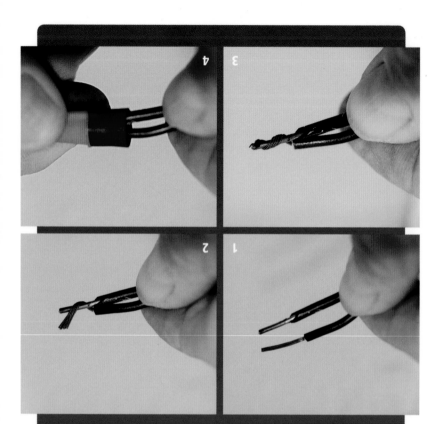

RETROFITTING LEDS

LED light sources can be installed in existing recessed lights (also called can lights, pot lights, or high-hats). If the recessed lights have incandescent or halogen bulbs in them, retrofitting them with LEDs will save a lot of energy. LED light sources for can lights save 80 percent of the energy that an incandescent bulb uses. Typical LED light sources also last 20,000 hours or more, so you'll be avoiding a lot of trips up and down a ladder to replace incandescent bulbs (which have a typical life of 2,000 to 3,000 hours).

LED Options

Some LEDs have a power source (ballast), an LED bulb (usually not replaceable), and an integral diffuser and trim ring. These come in different "color temperatures"; an LED with a 2,700K color temperature is very close to a regular incandescent bulb, providing a nice, warm light. An LED with a color temperature of 3,000K is a little on the cooler, or bluer, side. Some LEDs can be dimmed, with the right dimmer switch. You must buy and install a dimmer that is designed for LEDs. Most LEDs will fit 5-in.- and 6-in.-diameter can lights; check the LED spec sheet for compatibility with the can lights you have. Another option is to use an LED screw-based bulb, simply replacing the incandescent bulb.

LED LIGHT SOURCE. This is an integrated power supply, LED emitter, diffuser, and trim ring.

1. BULB FIRST. To retrofit an LED light source, switch off the power at the wall switch and the breaker. Unscrew the bulb from the incandescent fixture.

2. TRIM RING NEXT. Pull the trim ring down using your fingernails or a couple of putty knives. The baffle and trim will slide down a few inches.

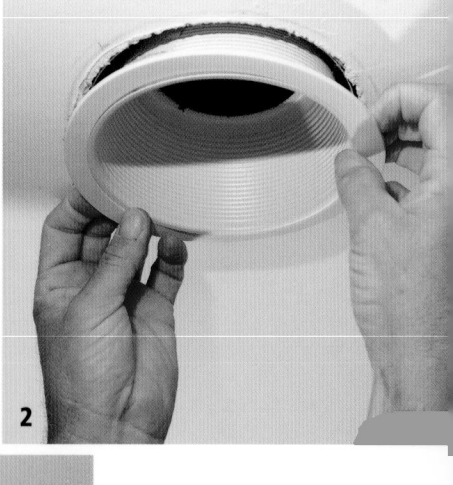

3. BAFFLE LAST. Slide the trim ring down farther, reach up into the fixture, and squeeze the retaining springs so that the baffle and trim come free. The recessed can stays put.

QUICK TIP Hold on to the retaining springs as you disassemble the fixture. If you slip, they'll whack your hand like a mousetrap.

4. PREPARE FOR RETROFIT. Screw the LED adapter into the socket in the can.

5. MAKE THE CONNECTION. Attach the LED to the adapter via the plug-in connector. Support the LED with your free hand so that it doesn't hang on the wires.

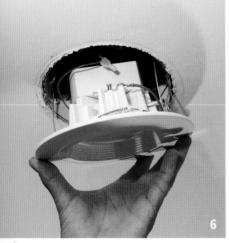

6. SLIDE INTO THE HOUSING. It may be held with retaining springs like the old baffle and trim (as shown here) or there may be spring tabs that you adjust to give a friction fit against the side of the housing.

STANLEY

REPLACING A SMOKE ALARM

Smoke alarms (also known as smoke detectors) save lives, giving you and your family a chance to get out of your house alive if there's a fire. Alarms that operate off both battery power and AC power are the most reliable. Alarms that are dual-powered and are interlinked are the best; if one unit sounds off, all units sound off.

To replace a smoke alarm, first shut off power to the circuit and test to verify. Then turn the body of the alarm and separate it from the mounting base. Remove the old mounting base and install the new one. If the unit is hard-wired and interconnected, you'll see a wiring adapter with black, white, and red wires connected to the building wiring. Remove the adapter and connect the new alarm's adapter. If the alarm is battery powered, install the battery or pull the plastic strip to connect the battery. Then, align the alarm body with the mounting plate, and twist it until it locks in place. Lastly, test the alarm. It's a good idea to use earplugs or other hearing protection, as you'll be within arm's length of an 85-decibel noise.

1. MAKE CONNECTIONS.
Match the wire colors from the adapter and splice to the building wire.

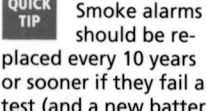 **QUICK TIP** Smoke alarms should be re-placed every 10 years or sooner if they fail a test (and a new battery doesn't fix it). Every unit is marked with a date of manufacture. Dispose of the old unit properly; check your local and state regulations.